What Happens in Spring?

Animals in Spring

by Jenny Fretland VanVoorst

Bullfrog Books

Ideas for Parents and Teachers

Bullfrog Books let children practice reading informational text at the earliest reading levels. Repetition, familiar words, and photo labels support early readers.

Before Reading
- Discuss the cover photo. What does it tell them?
- Look at the picture glossary together. Read and discuss the words.

Read the Book
- "Walk" through the book and look at the photos. Let the child ask questions. Point out the photo labels.
- Read the book to the child, or have him or her read independently.

After Reading
- Prompt the child to think more. Ask: What are the first animals you see in the spring? Where do you think they spent the winter?

Bullfrog Books are published by Jump!
5357 Penn Avenue South
Minneapolis, MN 55419
www.jumplibrary.com

Copyright © 2016 Jump! International copyright reserved in all countries. No part of this book may be reproduced in any form without written permission from the publisher.

Library of Congress Cataloging-in-Publication Data

Fretland VanVoorst, Jenny, 1972– author.
 Animals in spring / by Jenny Fretland VanVoorst.
  pages cm.—(What happens in spring?)
 Audience: Ages 5-8
 Audience: K to grade 3
 Includes index.
 ISBN 978-1-62031-234-6 (hardcover: alk. paper)
 ISBN 978-1-62031-478-4 (paperback)
 ISBN 978-1-62496-321-6 (ebook)
 1. Animal behavior—Juvenile literature.
 2. Animals—Juvenile literature.
 3. Spring—Juvenile literature. I. Title.
 QL751.5.F69 2015
 591.5—dc23

2014046778

Series Designer: Ellen Huber
Book Designer: Lindaanne Donohoe

Photo Credits: All photos by Shutterstock except: Dreamstime, 4, 23tl; iStock, 10; Thinkstock, 5.

Printed in the United States of America at Corporate Graphics in North Mankato, Minnesota.

Table of Contents

Life Is Good 4
Baby Animals of Spring 22
Picture Glossary 23
Index 24
To Learn More 24

Life Is Good

A groundhog leaves his burrow.

Is it time?

Yes. Winter is over.

A turtle crawls out of the water.

He spent all winter on the lake bottom.

Now it is time to eat.

Birds chatter in the trees.

They are back from the South.

It is time to find a mate.

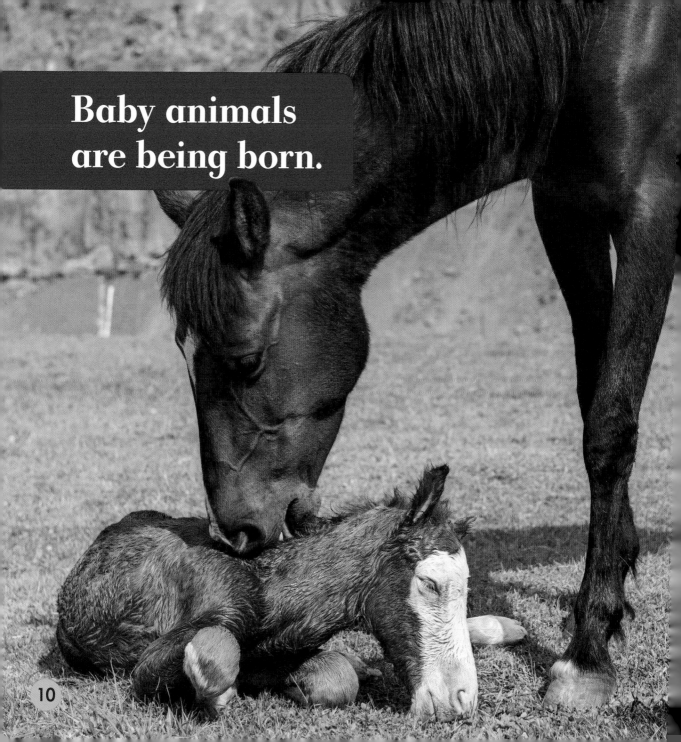

Baby animals are being born.

On the farm, a lamb learns to walk.

In the forest,
a bear cub
learns to climb.

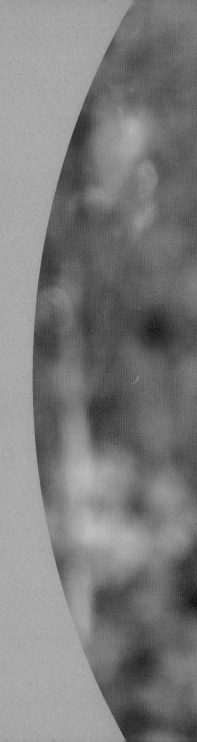

eggs

A frog lays her eggs in a pond.

Soon the eggs will be tadpoles.

Then they will be frogs.

tadpoles

There is so much to eat!

Grass and berries.

Fish and worms.

Life is good.

It is spring!

Baby Animals of Spring

Picture Glossary

burrow
A groundhog's underground home.

mate
A male or female partner of a pair of animals.

cub
A young mammal such as a bear, fox, or lion.

tadpole
A stage of a frog's life; tadpoles grow into frogs.

Index

babies 10
birds 8
cub 12
eating 7, 16
eggs 15
frog 15
groundhog 4
lamb 11
mate 8
South 8
tadpoles 15
turtle 7

To Learn More

Learning more is as easy as 1, 2, 3.
1) Go to www.factsurfer.com
2) Enter "animalsinspring" into the search box.
3) Click the "Surf" button to see a list of websites.